School-Home Connection

Grade 5

Harcourt Brace & Company

Orlando • Atlanta • Austin • Boston • San Francisco • Chicago • Dallas • New York • Toronto • London

http://www.hbschool.com

W9-AUF-247

Grateful acknowledgment is made to Suzanne Sutton for permission to reprint
from "Finding the Glory in the Struggle: Helping Our Students Thrive When
Math Gets Tough" by Suzanne Sutton in *The National Association of Secondary
School Principals Bulletin*, February 1997. Text copyright © 1997 by Suzanne
Sutton.

Printed in the United States of America

ISBN 0-15-311144-5

3 4 5 6 7 8 9 10 170 2000

Math Advantage

Parent Preview

This brochure provides information that can help your child achieve success using Math Advantage.

The older your child is, the more difficult it is to know what's happening at school. It's hard to help your child succeed in school unless you know what they are learning. This brochure is a preview of

- what your child will be learning in math this year,
- some of the vocabulary words that are important to know,
- and a few activities to do throughout the year that will be fun as well as good practice.

The **Math Advantage** program is also helpful in many other ways. Throughout the year you will receive **Math at Home** letters. Each letter will be sent to you at the beginning of each short chapter. It describes

- the specific mathematics concepts and vocabulary that your child will be learning,
- and additional activities to do at home that will provide extra practice.

By keeping these letters in a folder, you can help your child review and improve math skills all year long.

At the end of some chapters in your child's math book is a section called *Math Fun*. It has three activities to practice the math content just learned. One activity has an idea for extending the activity at home. Look for this symbol [🏠] on each *Math Fun* page.

"Why do I have to learn this?"

This is the question both parents and teachers hear from students. You can give your child a better understanding of how math is used in real life by sharing with them the ways in which you use math every day.

Point out how you use math to

- calculate coupon savings.
- check change given by a store clerk.
- count the hours, days, weeks, or months until a special occasion arrives.
- calculate how much money you will need to buy lunch for the family.
- decide how to double a recipe.
- budget time when running several errands.

Talking with your child about these routine events in life can lead to all kinds of exciting conversations. By sharing the value of math in this casual way, your child will also value what he or she is learning.

Local, State, and National Goals

The instruction and math content in **Math Advantage** is presented in a way designed to develop these local, state, and national goals. Students should

1. learn to value mathematics.
2. become confident in their ability to do mathematics.
3. become mathematical problem solvers.
4. learn to communicate mathematically.
5. learn to reason mathematically.

Students achieve these goals by learning the skills, following a thinking process, and using a variety of strategies to solve problems.

Page 1 in your child's math book, called *Be a Good Problem Solver*, describes a plan that will help your child successfully solve problems. Discuss the questions and suggestions given with this plan if your child needs extra help on a particular problem. Help your child use this process by giving prompts for thinking about and solving a problem.

Understand the Problem

Plan How to Solve It

Solve It

Look Back and Check Your Work

What Your Child Will Learn This Year

The core skills and some of the words that are developed in **Math Advantage** during the year are listed in the table below. Look at the Word Power boxes and the lessons in your child's math book for the way these words are used.

CORE SKILLS	VOCABULARY WORDS
understand numbers and place value of whole numbers and decimals	*thousandths, mixed decimals, equivalent decimals, equivalent fractions*
add, subtract, multiply, and divide whole numbers and decimals	*estimate, decimal point, product, quotient, divisibility*
organize, graph, and analyze data	*mean, median, mode, range, interval, scale, line graph, circle graph*
calculate probability	*certain, equally likely, impossible, tree diagram, possible outcome*
measure with customary and metric units	*centimeter, meter, millimeter, kiloliter, liter, milliliter, benchmark, volume, capacity, cubic units*
understand fractions and number theory	*numerator, denominator, mixed number, least common multiple (LCM), greatest common factor (GCF), simplest form*
add and subtract fractions and mixed numbers; multiply fractions	*least common denominator (LCD), like fractions, unlike fractions*
explore polygons and their movement in the coordinate plane	*transformations, congruence, symmetry, tessellation*
understand properties of a circle	*circumference, angles, radius*
recognize parts and properties of plane and solid figures	*nets, quadrilateral, isosceles, scalene, and equilateral triangles, chord, prism, base pyramid*
explore ratio and percent	*ratio, percent, benchmark percent*

ACTIVITIES

Introduce these activities when you know your child has studied the content. Repeat the activities throughout the year so that you can help your child review and improve his or her math skills.

Weather Watcher

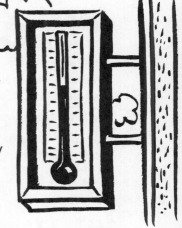

After completing this activity, your child will know more about weather in your area.

What you need: newspaper weather page

What you do:

1. Have your child cut out the daily weather page for one week.

2. Have your child find the following statistics:
   a. the difference between the high and low temperature each day (the range)
   b. the difference between the high temperature of the first day and the high temperature of the seventh day
   c. the difference between the high temperature in your city and the high temperature in the city of a relative or friend.

3. Ask your child to find the average high temperature for the seven-day period. He or she may do this by adding all the high temperatures and dividing the sum by 7 (the total number of days). Your child can find the average low temperature in the same way.

4. As an extension, you may wish to have your child list the *range* (part 2a) for each of the seven days. Then have your child find the *average range* of the temperatures. (Science note: during winter and summer, the high-low temperature range is usually smaller. During change of season, the high-low temperature range is usually greater.)

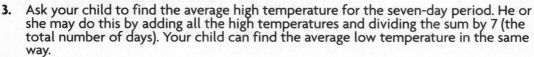

Operation: Shop

This activity focuses on estimation. It emphasizes to your child that mental math and estimation are a real part of daily life.

What you need: products sold singly and in packages
Note: You may wish to do this activity while shopping.

What you do:

1. Focus your child's attention on the use of addition, subtraction, multiplication, and division when they shop.

2. **Ask questions that highlight multiplication skills such as, "How much would two pairs of socks cost if one pair is about $3.00? How many pairs of socks could we buy for $10.00?"**

3. Questions that highlight subtraction skills involve comparison shopping. These include, **"Which pair of socks is less expensive? About how much less? How much change would we get from a $20 bill if we bought 3 pairs of socks?"**

4. Division concepts are used when answering questions about bulk packaging such as, **"About how much would one pair of socks cost if we bought each pair individually? Is it less expensive to buy them individually or in a package of 3 or 6?"**

5. Focus your child's attention on the skills he or she used to answer the questions. You may extend this activity by asking your child to make up questions for you to answer.

Perfect Timing

This activity is an application of skills involving addition and subtraction of time and elapsed time.

What you need: analog clock with minute hand or digital clock

What you do:

1. Have your child use a clock to calculate cooking times. For example, you might say, **"If a recipe calls for a baking time of 30 minutes and you put it into the oven at 4:20, what time should it be done?"**

2. Add **"what ifs"** to reinforce and extend the concept...."**What if you put it in the oven at 5:00? 5:45? What if the cooking time was 45 minutes? What if the cooking time was 2 hours?"**

3. For a more challenging extension, help your child work backwards. Say, for example, **"Dinner must be ready at 6:00 and the chicken takes $2\frac{1}{2}$ hours to roast. What time should the chicken go into the oven?"**

Perimeter and Area Hunt

In this observation activity, your child uses common household objects and situations to apply the concepts of perimeter and area.

What you need: no materials

What you do:

1. Have your child find a framed photograph or framed picture in your home. Say, **"Suppose you are framing this photograph. Would you measure around the outside of the photo or the area inside?"** (outside; the frame goes around the perimeter of the photo.)

2. Have your child look at the kitchen floor and say, **"Suppose you want to put new tile in this room. Would you measure around the outside or the area inside of the room?"** (inside; the tile will be laid in the inside area of the room.)

3. Walk around your home with your child, identifying other things for which a perimeter or area measurement is needed. For perimeter, you might identify window frames, door molding, and the edging around a blanket. For area, you might find carpeting, wallpaper, or bathroom tile.

4. As an extension, your child could use a measurement tool to find the actual perimeter or area of one or more of the items found on your hunt.

Internet Activities

You can access additional math activities on the Harcourt Brace Home Page. The address is **http://www.hbschool.com**

Volume 5   No. 1
http://www.hbschool.com

MATH AT HOME

Math: The Lighter Side

In a world in which large numbers are used to describe everything from the world's population to the national debt, it is important that a child have a good sense of number. Part of good number sense is a feeling for the size of numbers and how they relate to each other.

Place Value of Whole Numbers

Your child is beginning a chapter that will review and extend what he or she has learned about place value. In this chapter your child will study numbers through hundred millions, relationships among numbers, and ways large numbers are used in everyday life. Point out to your child examples of the use of large numbers, and discuss different ways numbers are written.

Here is an activity based on place value to do with your child.

Wipe Out

Have your child enter the number 9376524 into the calculator, and ask him or her to read the number to you. Explain that you will ask him or her to remove a given digit—for example, the 6—from the number display. The child may not clear the calculator and enter the new number to do this but must subtract the value of the digit—in this case, 6,000—from the number. Ask the child to remove one digit at a time until the entire number has been wiped out. Then start again with a new number.

HOMEWORK Tip

Establish a set time and a quiet place for your child to complete homework. Having a routine for homework helps him or her get organized and focused.

VOCABULARY

cardinal—numbers that tell how many
ordinal—numbers that tell position or order
nominal—numbers that name things

Math: The Lighter Side

The student in the cartoon thinks the right answer is the goal. Parents should not be fooled into thinking their child understands a math idea simply because he or she gives a correct answer to a problem or question. Parents should ask the child to explain how he or she got the answer.

Adding and Subtracting Whole Numbers

Your child is beginning a chapter that reviews addition and subtraction of whole numbers. Students will estimate sums and differences and use addition and subtraction in many situations. Share with your child some of the ways you use addition and subtraction in daily life. Give your child the chance to practice addition and subtraction skills in situations around the house.

Here is an activity you can do with your child to help him or her practice addition and subtraction.

Missing Digits

Find the digits that complete these problems to produce correct sums.

1.	347	**2.**	8△6	**3.**	8,△6△
	6△7		16△		−△,574
	+△98		+382		2,9△6
	1,902		△,△62		

Make up some problems of your own.

HOMEWORK @ Tip
Be sure your child completes assignments neatly and shows the work in an organized way. This will allow the child to check the work and the teacher to follow the solution.

TESTING @ Tip
When preparing for tests, your child should review basic facts of addition, subtraction, multiplication, and division on a daily basis. This will boost his or her confidence, accuracy, and speed on computation questions on the test.

Volume 5   No. 3
http://www.hbschool.com

MATH AT HOME

Place Value of Decimals

Your child is beginning a chapter that shows how decimals are related to fractions. Students will read, write, and use decimals to the thousandths place, find equal decimals, and compare decimals and put them in order. Share with your child ways you use decimals in your daily life.

Here is an activity to do with your child that will help him or her understand place value in decimals. The Extension challenges you and your child to find uses of decimals in the newspaper.

Make It Big!

Write the digits 0–9 on ten small slips of paper, one digit per slip. Place the slips in a small bag. Each player should make a gameboard on a sheet of paper with six blanks and a decimal point, as shown.

_ _ _ . _ _ _

Players shake the bag, and each pulls out a slip of paper without looking. They write their digit on one of their blanks. Once the digit is written it cannot be moved. Then they return the slip of paper to the bag and shake the bag. Then they each pull out another digit and write it on another blank. When the numbers are complete, players compare them. The player with the greater number wins.

Extension • Daily Decimals

With your child, look through the newspaper to find five different uses of decimals. Talk about how the numbers are used and what information they give.

Math in the News

What does it mean to say gas prices rose 4 thousandths of a dollar in a one week period? We use a lot of decimals in our daily life—for example, in scores, prices, and averages. To use decimals, we need an understanding of place value, and an ability to compare numbers and put them in order. Sharing with children ways decimals can be used helps them see that mathematics extends beyond the textbook.

Homework Tip

Encourage your child to try every problem, even those he or she finds difficult. Many times he or she will be able to get farther in the solution than expected.

Vocabulary

thousandth—one part of 1,000 equal parts
equivalent decimals—decimals that name the same number or amount
0.5 = 0.50 = 0.500

Possible answer: money amounts less than $1, budgets that involve large amounts of money $1.4 billion, a part of a school population 0.50, amounts of land or acreage 1.4 acres, cost of school lunch $2.15.

MATH AT HOME

Adding and Subtracting Decimals

Your child is beginning a chapter that takes adding and subtracting decimals to thousandths. Up to now, students have worked with decimals to the hundredths place and related them to money. Help your child to see the connection between adding and subtracting thousandths and adding and subtracting hundredths.

Here is an activity for you and your child that uses adding and subtracting decimals to solve magic squares.

It's Magic

A magic square is a square array of numbers in which the sum for any row, column, or diagonal in the square has the same value. For example, this is a 3 × 3 magic square with a sum of 15 for each row, column, and diagonal.

8	1	6
3	5	7
4	9	2

Complete the following 4 × 4 magic square. Be sure the numbers you fill in make rows, columns, and diagonals that have the same sum. The sum is called a magic sum. What is the magic sum?

8.806	A	B	11.322
C	16.354	D	E
F	10.064	12.58	18.87
17.612	13.838	G	5.032

It's Magic: A = 2.156, B = 20.128, C = 15.096, D = 3.774, E = 7.548, F = 1.258, G = 6.290.

For Your Information

Condensed from "Learning by Example," The Oregon Mathematics Teacher, Oct/Nov 1993

The ability to solve problems is learned by being with people who are solving problems. Here's how you can encourage your math student at home.

a. Set a good example. Let the child see you using mathematics. Talk about your thinking, about the steps you're following. Let the child see how important the skill of mathematics really is.

b. Talk and listen. The more math your child investigates, the easier it will be to solve problems. If your child believes math matters, he or she is more likely to continue using it.

c. Make math a focus. Help your child become stronger in math—solve math problems with him or her. Play games that use math skills. Talk about real-life problems.

HOMEWORK Tip Remind your child of the importance of carefully reading directions. He or she might be asked to find an answer, solve a problem, provide an estimate, or write an explanation.

TESTING Tip Remind your child to be careful when copying a problem from the test paper onto work or scrap paper. Often children write the problem down incorrectly, causing an error.

Volume 5 No. 5
http://www.hbschool.com

MATH AT HOME

Multiplying by One-Digit Numbers

Your child is beginning a chapter that shows how to multiply numbers by a one-digit number. Problems in the chapter teach key ideas of multiplication. They also show how multiplication is used to find area and volume. Give your child chances to practice this skill. If he or she has had difficulty with basic multiplication facts, be sure to review them.

The following activity and Extension focus on estimating products to help win a game.

Tic-Tac-Toe

Copy the grid below onto a sheet of paper along with the two factor banks. One player is X and the other O. X goes first. X chooses one number from each bank and multiplies the two to find their product. If the product is on the grid, he or she marks an X over it. Then it is O's turn. The first player to get 3 marks in a row wins. You may want to use a calculator to help with the multiplication or to check answers.

Bank 1		Bank 2		
7 9		237	96	294
		125	108	353

1125	2471	972
864	756	875
672	2646	2058

Extension • Your Move

Create your own tic-tac-toe game. Write 2 one-digit numbers for bank 1 and 6 two- or three-digit numbers for bank 2. Find the 12 possible products of multiplying one number from each bank. Write 9 of these products in the grid, and invite two other players to play!

Math Through Children's Literature

Children's books can provide important connections with mathematics. They often have ideas woven into their stories. This helps children see the use of an idea outside of the classroom.

Mitsumasa Anno has written many beautiful books that show math ideas through interesting problems for the reader. *Anno's Magic Seeds* (Philomel Books, 1995) is based on patterns, multiples, and doubling. Your child might enjoy reading about a boy who is given two magic seeds, one to eat and one to plant. He is promised that he will not go hungry for the second seed will grow two more seeds. But what happens when the boy decides to plant both seeds?

HOMEWORK @ Tip

Be sure your child reviews problems he or she missed earlier.

VOCABULARY

Commutative Property of Multiplication—factors can be multiplied in any order without changing the product

Associative Property of Multiplication—factors can be grouped in different ways without changing the product

Property of One—when one of the factors is 1, the product equals the other factor

Zero Property for Multiplication—when one factor is 0, the product is 0

MATH AT HOME

Multiplying Larger Numbers

Your child is beginning a chapter that teaches more about multiplication of whole numbers. In this chapter your child will learn to multiply by two- and three-digit numbers and to use multiplication for finding the area of figures. Give your child chances to practice these skills, and share some of the ways you use multiplication in daily life.

In the activity that follows, you and your child will estimate the results of multiplication. The Extension looks for patterns in multiplication.

Hit the Target

In each problem, choose a number by which to multiply the given factor that will make the product fall in the range. Check by doing the multiplication with paper and pencil or a calculator. Will more than one factor work for any of the problems?

$$682 \times \triangle \rightarrow \begin{bmatrix} 12,000 \\ \\ 11,000 \end{bmatrix}$$

$$789 \times \triangle \rightarrow \begin{bmatrix} 21,000 \\ \\ 20,000 \end{bmatrix}$$

$$298 \times \triangle \rightarrow \begin{bmatrix} 10,000 \\ \\ 9,000 \end{bmatrix}$$

$$302 \times \triangle \rightarrow \begin{bmatrix} 12,250 \\ \\ 12,000 \end{bmatrix}$$

Extension • Follow the Leader

Find the first three products, and then predict the next three, based on the pattern formed. Check by doing the multiplication. Do the factors 101 and 404 also form patterns?

303×5	303×10	303×15
303×20	303×25	303×30

202×5	202×10	202×15
202×20	202×25	202×30

For Your Information

Your child will use many mathematics tools, including a protractor, a compass, a ruler, and a calculator. The calculator has become a key tool in mathematics education.

This does not mean that a child need no longer know basic facts or do paper-and-pencil work. Rather, the calculator allows students to explore difficult problems, and discover how numbers are related.

As with any tool, a child needs to know how and when to use the calculator. The *how* will depend on what the calculator can do and on the age of the child. The *when* will depend on the situation. Children need to know that there are many times when it is quicker and easier to complete a problem without a calculator. Also, the tool is only as good as the data put into it, so children need to have good estimation and number sense skills.

TESTING @ Tip
Remind your child that checking for reasonableness of answers is important when taking a test.

VOCABULARY

Distributive Property—
multiplying a sum by a number is the same as multiplying each addend by the number and then adding the products
$10 \times 15 = 10 \times (10 + 5) =$
$(10 \times 10) + (10 \times 5)$

Hit the Target: 682 × 17, 298 × 31–33, 789 × 26, 302 × 40
Extension: 1,515; 3,030; 4,545; 6,060; 7,575; 9,090; 1,010; 2,020; 3,030; 4,040;5,050; 6,060; yes

6

MATH AT HOME

PEANUTS®

Reprinted by permission of United Feature Syndicate.

Math: The Lighter Side

As Marcie in the cartoon knows, children should study for a math test. To do this they should:
- practice solving problems like those in the unit the test is on.
- review homework assignments, paying attention to problems that were incorrect.
- review math vocabulary words.
- go over the chapter review.

Dividing by One-Digit Numbers

Your child is beginning a chapter that reviews dividing by a one-digit number. Students will focus on divisibility rules and compatible numbers. They will also learn the role of zero in the quotient, find the remainder in problems, and practice division.

Here are an activity and an Extension for you and your child to try that use divisibility rules to solve riddles and find missing digits.

Division Mix-Up

Form a number with the digits given in each problem so that the stated condition is met. Each digit can be used only once in each number.

a. Use the digits 5, 4, 0 to form a number that is divisible by 6.

b. Use the digits 4, 2, 6 to form a number that is divisible by 8.

c. Use the digits 4, 5, 9, 3 to form a number that is divisible by 6.

d. Use the digits 4, 3, 8, 1 to form a number that is divisible by 7.

Extension • Lost Digits

Replace the missing digits in each number so that the resulting number will be divisible by the given number. Make up some more of your own to ask each other.

$$83,76\triangle \div 6$$
$$38,\triangle52 \div 9$$
$$1,876,9\triangle2 \div 4$$

VOCABULARY

divisible—a number is divisible by another number if the quotient is a whole number and there is no remainder

compatible numbers—numbers close to the actual numbers that can be divided evenly
$408 \div 4 = n$ Think: $400 \div 4 = 100$
n is approximately (~) 100
400 and 4 are compatible numbers

Dividing by Two-Digit Numbers

Your child is beginning a chapter on long division with a two-digit divisor. Helping your child practice using estimation in division will increase his or her success in this chapter.

Here is an activity for you to do with your child that will provide extra practice with division. The Extension presents a puzzle for you and your child to solve. Good luck!

Deal Them Out!

For this activity you need a deck of regular playing cards. Use only the cards ace (1) to 9. The cards will be used to stand for the digits in the divisor and dividend of a problem.

Deal out three cards. Use the three cards, in the order dealt, to form a three-digit dividend. Then deal out two more cards. Use the cards, in the order dealt, to form a two-digit divisor. Have your child complete the division problem they form. Use a calculator to complete the multiplication check for the problem.

Example: | 6 | 4 | 7 | 2 | 3 |

$$\begin{array}{r} 28 \text{ r}3 \\ 23\overline{)647} \\ -46 \\ \hline 187 \\ -184 \\ \hline 3 \end{array}$$

Check: $(28 \times 23) + 3 = 647$ ✓

Extension • Missing Digits

Fill in the missing digits in the division problem. X need not represent the same digit.

$$\begin{array}{r} 25 \\ 5X\overline{)1,25X} \\ -1\ 00 \\ \hline 250 \\ -2XX \end{array}$$

For Your Information

Writing has become a key part of the study of mathematics today. Parents who never had to write to learn mathematics might feel confused by this. Why is writing in math important?

First, asking children to write about what they are doing gets them to focus on understanding the problem and knowing how to solve it rather than just on getting the right answer.

Second, writing helps children see connections between math and other study areas. Students become aware that mathematics is more than just strings of numbers. And last, having to explain to someone what they have done and why, helps children better understand these things themselves.

Educators are finding that the pen is a powerful key to unlock student thinking.

HOMEWORK Tip Look over homework assignments to be sure your child is reading and writing notation correctly. $29\overline{)364}$ must be read as "364 divided by 29," not "29 divided by 364."

TESTING Tip Remind your child to carefully review homework assignments in preparation for a test. The problems and examples in the assignments show what will be expected on the test.

Volume 5   No. 9
http://www.hbschool.com

Analyzing and Graphing Data

Your child is beginning a chapter on dealing with data. Problems in this chapter will use mean, median, and mode. Students will learn to choose and make the right graph for the kind of data to be shown. Share with your child any examples of graphs you can find. Understanding data and the ways it can be shown are important skills in today's world.

Here is an activity that looks at how a median can describe data. The median is the middle value of a set of data arranged from least to greatest.

Middle of the Road

For this activity you and your child will need to record the temperature in a city for seven days in a row. You can use the temperature in any city listed in the newspaper, or you can record the temperature in your own town for seven days. At the end of the seven days, write the temperatures in order from least to greatest. Find the median, or middle value, for the set of data.

Now you can experiment with the median. Start with the complete set of data each time.

a. Remove two temperatures without changing the median.

b. Remove two temperatures and increase the median.

c. Remove two temperatures and decrease the median.

d. Add two temperatures that will increase the median. Then add two that will decrease it.

e. Add two temperatures that will not change the median.

Possible answer: numbers in a series are often compared to the median in relation to how much above or below the middle point they fall—game scores, test scores are examples; 0, 1, 500

For Your Information

Condensed from Guidelines for The Teaching of Statistics, Center for Statistical Education, American Statistical Association

"**W**e live in a world of information, in which numbers are all around us. They affect our decisions on health, citizenship, parenthood, employment, sports, and money matters. To be an informed citizen today, a person must be able to deal with this data. Because of this, American schools must teach students more about statistics and probability."

A basic understanding of statistics and probability will be needed by all who live in the twenty-first century. Parents need to help children understand these math ideas and use them in their lives.

HOMEWORK @ Tip

In homework problems dealing with graphing, make sure your child identifies and labels the title and the graph. Also, check whether he or she has chosen a scale that is accurate and that describes the data.

VOCABULARY

median—the middle number in an ordered series of numbers

mode—the number that occurs most often in a list of data (Example: The mode of 1, 3, 4, 4, 6 is 4.)

mean—one way to find a number that represents all numbers in a set

scale—the series of numbers placed at fixed distances on a graph

interval—the distance between the numbers on the scale of a graph

MATH AT HOME

Circle, Bar, and Line Graphs

Your child is beginning a chapter on graphs to show data. Students will use sections of a circle graph to show fractions or decimals. They will also read and make, and compare graphs. Share with your child examples of graphs you find, and discuss the purpose of each graph with your child.

Here are an activity and an Extension to do with your child that will provide practice in making and reading circle graphs.

You're on Candid Camera

In this activity you will collect data that relate the number of minutes of television program time to the number of minutes of commercial time. Choose a favorite one-hour television show. While you watch it, keep track of the number of minutes of commercials that are shown. Have your child draw a circle graph that shows the relationship between the program time and the commercial time. To make this easier, draw the graph as a clock. It will make it easier to mark off the number of minutes of commercial time. Show the two areas of the graph as both fractions and decimals.

Extension • Prime Time T.V.

Does the number of minutes of commercial time change with the time of the day or the type of program being shown? Repeat the activity above, viewing a program in the morning, in the late afternoon, and in prime time (8–11 p.m.). Compare the amounts of time given to commercials. What do you discover?

Math in the News

Condensed from Associated Press News Release

This shows what children can do when they use mathematics to check out a claim.

Third grade pupils at Wadesboro Central Elementary School decided to sharpen their math skills recently by checking out an advertising claim that there are 1,000 chocolate chips in each 18-ounce bag of Chips Ahoy! They made a startling discovery—instead of 1,000, they found from 340 to 680 chips per bag. Nabisco sent a team of "cookie technicians" to the school to explain why the youngsters came up short. Their explanation was that "sometimes the chips get smashed in the baking process, and sometimes the chips fall apart."

TESTING Tip

Remind your child that before beginning a test, he or she should look over the entire test to decide how best to complete it. He or she might want to do problems or answer questions that seem easier before doing those that might take more time.

VOCABULARY

double-bar graph—a graph that uses two sets of bars to compare two sets of data

line graph—a graph that uses a line to show how something changes over a period of time

line plot—a diagram that shows the frequency of data as they are collected

Volume 5   No. 11
http://www.hbschool.com

MATH AT HOME

B.C.   **by johnny hart**

Reprinted by permission of Johnny Hart and Creators Syndicate, Inc.

Math: The Lighter Side

The cartoon suggests that successful math students would know that someone cannot be "with you 110%"—100% would be the highest amount possible. In the study of probability, students learn that when an event is absolutely certain to occur, the probability is 1, or 100%. Also, when it is not possible for an event to occur, the probability is zero.

Discuss with your child examples of situations that are certain, impossible, highly likely, somewhat likely, and unlikely.

Probability

Your child is beginning a new chapter that explores and develops basic concepts of probability—the chance of an event happening. Problems in the chapter involve identifying events that are certain, impossible, or likely to happen; determining outcomes for events; and finding and comparing probabilities. Share with your child applications of probability that you encounter in daily life.

Here is an activity for you and your child to do together that explores the expected probabilities of an event happening.

Let Them Roll

For this activity you will need two number cubes. Roll the number cubes and add the two numbers. What number did you get for the sum? Roll again. Did you get a different sum? Identify what sums are possible when two number cubes are rolled.

How can you obtain a sum of 5 when rolling two number cubes? Is there another sum that can be obtained in the same number of ways as 5? Are there sums that are more likely to occur than the sum of 5? Are there sums that are less likely to occur than the sum of 5? How would you describe being able to roll a sum of 1? How would you describe being able to roll a sum of less than 13?

HOMEWORK @ Tip

Be sure to ask your child to summarize the main points of a homework assignment. This should help the child reflect on key concepts.

VOCABULARY

equally likely—when the outcomes of an experiment have the same chance of happening

event—the action that happens in an experiment that brings about an outcome

probability—the chance that a given event will occur

MATH AT HOME

Multiplying Decimals

Your child is beginning a chapter on multiplying decimals. Share with your child some of the ways you use decimals every day.

Here is an activity in which you and your child can have fun finding decimals in the newspaper. The Extension provides practice in adding and multiplying decimals.

Newspaper Scavenger Hunt

How many of these decimal numbers can you find in the paper? Cut them out, and paste them on a sheet of colored paper to make a collage. You may find other categories of decimal numbers not on this list!

1. a number less than 1
2. a metric measurement
3. a scale on a bar graph
4. a number expressed to hundredths
5. a decimal written in words
6. an average
7. an estimate
8. a number greater than 1 million
9. a sport statistic expressed as a decimal
10. time expressed as a decimal

Extension • What's a Word Worth?

Assign money values to each of the letters of the alphabet, for example:

A = 1¢ = 0.01   B = 2¢ = 0.02   C = 3¢ = 0.03

Then find as many words as you can in the newspaper that are worth $1.00. For example: The word "pupils" is equal to 0.93. You add or multiply each letter's value.

Math in the News

Condensed from an Associated Press article by Jennifer Goodwin

It's easy for children to see their parents using reading skills but not so easy to see them using math skills, as this is often done mentally. Parents need to show children how they use math in daily life.

Parents can visit the library for books of math activities and games to use at home. These can show children that math is in everything they do.

Create a Homework Box of supplies so your child will feel "ready to go." Include such things as paper, a ruler, colored markers or crayons, sharpened pencils, graph paper, and an eraser.

mixed decimal—a number that has a whole number part and a decimal part

Volume 5   No. 13
http://www.hbschool.com

MATH AT HOME

Dividing Decimals

Your child is beginning a chapter on division of decimals. Students will need to estimate quotients and divide a decimal by a whole-number divisor. Share with your child times when you use decimal division, as, for example, with money.

Here is an activity to do with your child that uses division of decimals to find an average. The Extension explores the idea of unit pricing.

It's a Handful

For this activity you will need a large collection of pennies, nickels, dimes, and quarters. Place the collection in a large bowl, and be sure the coins are well mixed. Have your child take a handful of coins and sort and count them to see how much money he or she scooped up. Have the child return the coins to the bowl, mix them in, and repeat five times. Compare the value of the coins scooped up in each of the five turns. What was the greatest amount? What was the least? Find the average amount of money scooped up in the five turns. You may wish to take turns with your child and compare your results.

Extension • One at a Time

Share with your child how unit pricing is used in a supermarket. Using a food ad from the newspaper, find an example of a product sold in multiple quantities, for example, 3 cans of juice for $2.59. Have your child find the cost of buying one. Show your child the shelf label in the supermarket that gives the unit price, or cost for one.

For Your Information

Estimation—giving an answer close to the exact answer when an exact answer is not needed—is an important part of your child's study of mathematics.

Children are taught a number of strategies to use to estimate. One of the most commonly used strategies is rounding. Often children are asked to estimate the answer before solving the problem. This is done so that the child will be able to tell if the answer he or she gets is reasonable. Children often resist doing this step, seeing it as unnecessary. Parents need to help their children understand the purpose of estimation and encourage them to use it.

HOMEWORK @ Tip Remind your child that the goal of homework should be more than just getting the assignment done. Children should look for ways the problems relate to each other. They should also be sure they understand the steps used to get the answers.

TESTING @ Tip Remind your child to use inverse (opposite) operations to check answers to problems on a test. For example, when dividing decimals, he or she should use multiplication to check the quotient.

MATH AT HOME

Measurement: Metric Units

Your child is beginning a chapter on metric measure. Students will learn how to measure length, mass, and capacity in metric units. They will also learn how units of metric measure are related, how to change units of metric measure, and how to solve problems using metric measure. Help your child practice metric measurement at home. Point out the metric measurements on packages you buy.

Here is an activity to do with your child that uses measurement to solve problems. The Extension challenges you to find objects of a given size.

Making Cents of Metrics

For this activity you will need pennies, nickels, dimes, and quarters and a metric ruler. Set up the situations in the questions and measure the thicknesses and diameters (distance across) of the coins to find the answers.

1. Which is worth more, a line of quarters 20 cm long lying on the table, each touching the next, or a stack of nickels 10 cm tall?

2. Which is worth more, your height shown as a stack of nickels or as a line of quarters?

3. A dime weighs about 2 grams. Which is worth more, 1 kg of dimes or a line of quarters 1 meter long?

Extension • Seek and Find

Find five different objects in the house that are about 1 centimeter in length and five objects that are 1 meter in length.

For Your Information

Have you ever had to estimate how long or how heavy something is? Before you can make a good estimate, or "educated guess," you have to have a sense of commonly used measures. Before you can tell if a room is 9 feet long, you need a sense of how long a foot is.

You can help your child develop a good measurement sense by teaching him or her a set of personal reference marks to compare objects to.

Here are a few commonly used references for adults. You and your child can make a list that fits his or her size.

one inch = the distance between the knuckles on your middle finger
one foot = the length of your foot
one yard = the distance from the tip of your nose to the tip of your finger with your arm stretched out
one centimeter = width of your little finger
one meter = a little longer than a yard

TESTING Tip Remind your child to label all measures with the units used, both in answers and in diagrams.

VOCABULARY

meter (m)—a unit of length in the metric system
gram (g)—a unit of mass in the metric system
liter (L)—a unit for measuring capacity in the metric system

Making Cents of Metrics: 1. both equal to about $2.00 **2.** stack of nickels **3.** kg of dimes

MATH AT HOME

Understanding Fractions

Your child is beginning a chapter that teaches more about the fractions met in earlier grades. Problems in this chapter deal with fractions as part of a whole and part of a group. Students will read and write fractions, mixed numbers, and equivalent fractions. They will compare fractions and put them in order. Share with your child ways you use fractions, especially in measurement. The more experiences your child has with fractions, the more at home with them he or she will be.

Here is an activity about comparing fractions. The Extension is the reverse of work covered in the chapter.

Ready, Aim, Pick!

On slips of scrap paper, write the following fractions, one to a "card."

$\frac{1}{2}, \frac{2}{4}, \frac{3}{4}, \frac{5}{4}, 1\frac{1}{4}, \frac{6}{4}, \frac{2}{3}, \frac{1}{6}, \frac{3}{6}, \frac{4}{6}, \frac{5}{6}, \frac{7}{6}, 1\frac{5}{6}, \frac{4}{8}, \frac{5}{8}, \frac{6}{8}, 1\frac{3}{8}, 1\frac{7}{8}, \frac{2}{8}, \frac{12}{8}$.

Shuffle the cards and place them face down on the table in rows of 5. Each person picks up a card. The person with the greater value takes the pair. If the two cards have the same value, another set is drawn, and the person with the greater value takes both sets. The player with the most cards at the end wins.

Extension • What's the Whole?

Show what the whole would look like if

a. $\triangleright$ = $\frac{1}{4}$.　　b. $\vcenter{\hbox{:::}}$ = $\frac{2}{3}$.　　c. $\boxplus$ = $\frac{3}{2}$.

Make up some of your own!

Math in the News

What strikes your eye when you look at this ad? Is it the prices? Is it the beauty of the bracelet? Or is it all the numbers in the ad? You can use this ad to show your child how mathematics comes into our daily lives. Talk about the different uses of measurement in the ad—for wrist size and the width of the bracelet. Ask your child about how wide a bracelet measuring $\frac{7}{16}$" would be. What's another name for the width $\frac{4}{16}$"? Why did the ad writers choose not to write the $\frac{4}{16}$" in lowest terms? Continue asking questions about the ad.

Each night your child should spend five minutes practicing basic facts.

mixed number—a number made up of a whole number and a fraction
multiple—the product of 2 or more numbers

CALVIN & HOBBES

Reprinted by permission of Universal Press Syndicate.
© 1995 Universal Press Syndicate.

Math: The Lighter Side

Mathematics is going to be important to your child's career choice. Many more careers today need mathematical skills and understanding. Point out to your child some ways mathematics is used in different jobs.

Fractions and Number Theory

Your child is beginning a chapter that adds to what students learned about fractions in earlier grades. He or she will define prime and composite numbers, work with greatest common factors, find factors of numbers, give equivalent fractions, and write fractions in simplest form (lowest terms). Invite your child to share with you problems and activities done in class.

Numbers from Primes

A long time ago a famous mathematician named Goldbach made a conjecture about prime numbers. The conjecture claims that

* any even number greater than or equal to 4 can be written as the sum of two or more primes. (ex. 18 = 11 + 7)

* any odd number greater than or equal to 7 can be written as the sum of three or more primes. (ex. 15 = 3 + 5 + 7)

Pick at least 5 even numbers less than 100 and 5 odd numbers less than 100. Make a list of prime numbers you can use to test Goldbach's Conjecture.

Extension • Put It in Reverse!

23 is a prime number. When its digits are reversed, the new number, 32, is not prime. Your job is to find all the two-digit prime numbers that are also prime numbers when their digits are reversed. There are 11.

Do you see any patterns?

TESTING Tip

Remind your child that a test question may have more than one correct answer.

VOCABULARY

prime number—a number that has only 2 factors—itself and 1
composite number—a number that has more than 2 factors
equivalent fractions—fractions that name the same amount

Extension: 11, 13, 31, 17, 71, 19, 91, 37, 73, 79, 97

Volume 5   No. 17
http://www.hbschool.com

MATH AT HOME

Modeling Addition of Fractions

Your child is beginning a chapter that continues the work with fractions begun in the last chapter. Problems in this chapter include addition of fractions with like and unlike denominators. Share with your child examples of addition of fractions from daily life. These might include fractions used in cooking or building.

Here is an activity for you to do with your child that focuses on the sums of fractions. The Extension explores the use of unit fractions to express other fractions.

Fraction Puzzlers

Choose fractions from the box that fit each rule. Use number sense to help you choose your numbers.

$$\frac{1}{2}, \frac{2}{3}, \frac{3}{4}, \frac{1}{3}, \frac{1}{6}, \frac{5}{6}, \frac{1}{8}, \frac{3}{8}, \frac{5}{8}, \frac{7}{8}$$

*Find two fractions whose sum is less than $\frac{1}{2}$.
*Find two fractions whose sum is greater than 1 but less than $1\frac{1}{2}$.
*Find two fractions whose sum is $\frac{1}{2}$.
*Find two fractions whose sum is between $\frac{1}{2}$ and 1.
*Find three fractions whose sum is 2.

Extension • Fraction Building Blocks

A unit fraction is one in which the numerator is 1. Try to write any fraction between 0 and 1 as the sum of two or more different unit fractions. Here are two examples for you.

$$\frac{3}{4} = \frac{1}{2} + \frac{1}{4}$$
$$\frac{2}{5} = \frac{1}{3} + \frac{1}{15}$$

Can you express $\frac{3}{8}$ or $\frac{5}{6}$ using unit fractions?

For Your Information

It is important to help children develop good number sense, or common sense about numbers. In the primary grades most number sense activities focus on whole numbers (0, 1, 2, 3, 4, . . .).

In the intermediate grades, however, children must develop a number sense for decimals and fractions. For years, your child has learned that 8 is greater than 4. Now he or she must understand that $\frac{1}{8}$ is smaller than $\frac{1}{4}$. This confuses many children. Soon they must learn that multiplication of fractions does not produce a larger product and that division of fractions may not produce a smaller quotient. These new ideas can cause problems for children who do not have a strong number sense.

Parents can help children gain a greater number sense for fractions by finding ways for their children to use them in real life.

HOMEWORK Tip

Remind your child to use models to help find the answers to homework problems. For example, using fraction strips can help students find sums of fractions.

VOCABULARY

least common denominator (LCD)—the least common multiple of two or more denominators

$$\frac{3}{4} + \frac{2}{3} = \frac{9}{12} + \frac{8}{12}$$ twelfths is the LCD

Fraction Puzzlers: $\frac{1}{8} + \frac{1}{6}$; $\frac{2}{3} + \frac{1}{2}$; $\frac{3}{8} + \frac{1}{8}$; $\frac{1}{3} + \frac{1}{2}$; $\frac{3}{4} + \frac{7}{8} + \frac{3}{8}$.
Extension: $\frac{3}{8} = \frac{1}{4} + \frac{1}{8}$; $\frac{5}{6} = \frac{1}{3} + \frac{1}{2}$.

MATH AT HOME

Modeling Subtraction of Fractions

Your child is beginning a chapter on subtracting fractions. The chapter begins with how to do this and goes on to show uses for the skill, for example, in measurement.

Here are an activity and an Extension you can do with your child.

What's the Difference?

In early times Egyptians wrote all fractions as sums of unit fractions. Fractions such as $\frac{1}{2}$, $\frac{1}{3}$, $\frac{1}{4}$, $\frac{1}{5}$, and $\frac{1}{6}$ are called unit fractions because the numerator in each fraction is a 1, or a unit.

Write three subtraction problems in which the difference, written in simplest form, is a unit fraction. For example, $\frac{3}{4} - \frac{1}{4} = \frac{1}{2}$.

Extension • Make a Difference

Based on what you learned from the activity above, what two fractions would have a difference of $\frac{1}{6}$, $\frac{1}{2}$, or $\frac{1}{4}$?

For Your Information

Condensed from "Finding the Glory in the Struggle: Helping Our Students Think When Math Gets Tough," by Suzanne Sutton, NASSP Bulletin, Feb. 1997

Does your child struggle in math? What can you do to help?

Struggling is not the enemy in mathematics, any more than sweating is the enemy in basketball. It is a clear sign of being in the game. Math asks students to think in ways they are not used to. They are asked to look at familiar things in new ways, and then to use these ways to explore the unfamiliar things.

In their struggle to understand, and in the way they meet this struggle, they can learn life skills for use outside of the classroom. Mathematics offers them a chance to learn how to work through struggle. They learn how to bring to it what they have and how to find and use the things they need.

HOMEWORK Tip

Remind your child to use inverse operations to check answers. When finding the difference of fractions, for example, answers can be checked through addition.

TESTING Tip

Remind your child to use the textbook to prepare for a test. He or she should review bold phrases, worked examples, and definitions.

What's The Difference? Possible answers: 1. $\frac{3}{8} - \frac{1}{4} = \frac{1}{8}$; 2. $\frac{7}{12} - \frac{1}{4} = \frac{1}{3}$; 3. $\frac{5}{6} - \frac{1}{3} = \frac{1}{2}$.

MATH AT HOME

Adding and Subtracting Fractions

Your child is beginning a chapter that continues the study of fractions begun earlier. Problems in the chapter include finding sums and differences of fractions and connecting fractions to measurement on a ruler and to notes in music. Students will use their fractions skills in problem solving. Continue to share with your child ways you use fractions in your life.

Here are an activity and an Extension that use dominoes to practice fractions.

Fractions: Domino Style!

For this activity, use a set of dominoes from which all the doubles and blanks have been removed. Show your child that a domino can be thought of as a fraction less than one.

Place the dominoes face down on the table and mix them up. Roll a number cube. If the number rolled is even, pick up two dominoes and find their sum. If the number rolled is odd, pick up two dominoes and find their difference. Continue until all dominoes have been used.

Extension • It's a Ten!

Find 5 dominoes from the set used above that can be added to make a sum of 10. You can turn the dominoes to form fractions greater than 1 for this activity.
For example:

$$\boxed{\begin{smallmatrix} \cdot\, \cdot \\ \cdot\,\cdot\,\cdot \end{smallmatrix}} = \frac{3}{4} \quad \text{or} \quad \boxed{\begin{smallmatrix} \cdot\,\cdot\,\cdot \\ \cdot\, \cdot \end{smallmatrix}} = \frac{4}{3}$$

For Your Information

L L L		F E E T
I I I	FRACTION	E E
N N N		E E
E E E	FRACTION	T E E F
S S S		

Each of these signs stands for an idea from mathematics. Can you figure out what each represents? Sometimes problems have to be looked at in a different way to be solved. The ability to think creatively is a great help when it comes to solving problems. Problem solving is a main focus of the study of mathematics. Parents can help their children succeed in this area by encouraging their thinking skills. Logic problems, number puzzles, riddles, and strategy games all help develop thinking skills. The more children work and play with numbers, the better problem solvers they will become. Did you figure out that the signs above represent parallel lines, square feet, and writing fractions in simplest form?

HOMEWORK @ Tip
Your child should keep a math notebook. In it your child should copy some of the problems worked out in class as well as notes on explanations of math ideas.

TESTING @ Tip
Remind your child to be sure that answers on a test are simplified.

Extension: $\frac{1}{8} + \frac{3}{6} + \frac{1}{3} + \frac{2}{6} + \frac{3}{3} = 10$

Adding and Subtracting Mixed Numbers

Your child is beginning a chapter on adding and subtracting mixed numbers. Connections are made with earlier work done with fractions. Ask your child to share with you some of the problems done in class on this topic.

Here is an activity for you and your child that uses the skills of adding and subtracting mixed numbers to form patterns. A pattern is an ordered list of numbers that follows a rule. The Extension asks you to write mixed number sentences that equal a given value.

Play by the Rules

Start with the number given and keep adding the number that follows. Adding that number is the rule that will form a pattern. Add four numbers to each pattern. Be sure all numbers are whole numbers or mixed numbers in simplest form.

a. Start with $1\frac{3}{4}$; add $2\frac{1}{2}$.

b. Start with $2\frac{3}{8}$; add $1\frac{3}{4}$.

c. Start with 10, subtract $2\frac{1}{3}$.

d   Start with $18\frac{1}{2}$; subtract $1\frac{3}{4}$.

Extension • Same Solution

Write a fraction or mixed number in each box to make a true statement.

$\Box + \Box = 5\frac{1}{2}$

$\Box - \Box = 5\frac{1}{4}$

For each problem find three different pairs of numbers that satisfy the sentence.

For Your Information

Why do children learn so many different skills in their study of mathematics? Do they learn division simply to find quotients to rows of problems? Do they learn to add fractions merely to find sums of fractions? Do they learn to find the area of a rectangle just to practice multiplying length times width? Of course not!

No one studies mathematics just to learn a set of skills. Children learn these skills to be able to use them in real-life situations. So it is important to be sure that children see connections between what they are learning and how it will be useful to them.

Parents can help with this part of the job. For example, when children are studying fractions, parents can share ways they use fractions when measuring for building things or cooking.

Play by the Rules: a. $4\frac{1}{4}$ **b.** $4\frac{1}{8}$ **c.** $7\frac{2}{3}$ **d.** $16\frac{3}{4}$

Extension: Possible answers: $2\frac{1}{4} + 3\frac{1}{4}$, $10\frac{3}{8} - 5\frac{1}{8}$

MATH AT HOME

Measurement: Customary Units

Your child is beginning a chapter on measuring with customary units. Activities in the chapter focus on finding precise measures; using the right customary unit for length, capacity, and weight; changing customary units; computing with customary measures; elapsed time; and temperature change. Share with your child some of the uses of measurement found around the house.

Here is an activity for you and your child that involves measuring lengths in customary units.

Softball Toss

In the 1988 Olympics, shot-putter Uef Timmermann of East Germany tossed the shot 73 feet 8.75 inches. The men's shot weighs 16 pounds.

a.   Find a distance that measures about 75 feet.

b.   Find an object that weighs about 16 Ibs.

Do you think you would be able to throw that object that far?

Try the following with a standard softball. (A softball is a lot lighter than 16 lbs!)

Measure how far you and your child can toss the softball first overhand and then underhand. Compare the distances. It would be good to do this activity at a ball field if you can.

Extension • Tool Time

A ruler, scale, and cup are not the only measuring tools. Find out what each of the following instruments measures and in what units.

anemometer	galvanometer
barometer	sextant
seismograph	voltmeter
fathometer	

Math in the News

On DISN, one movie starts at 8:40 and the next one starts at 10:25. How much time is between the first movie and the second movie?

Schedules, whether they are for buses, trains, planes, television, or movies, deal with elapsed time. Children need to learn how to use schedules. They make math ideas come to life outside the classroom.

Children should be able to read the start and end times of TV shows and figure out the length of the show. Share with your child examples of elapsed time. This will support the activities done in school.

HOMEWORK @ Tip

Be sure your child properly uses measurement tools, especially rulers. Check to see that your child is placing the ruler properly to measure and reading the measure as asked. (For example: to the nearest $\frac{1}{4}$ inch)

VOCABULARY

precise measurement—finding the unit closest to the actual length of an object

degree Fahrenheit (°F)—customary unit for measuring temperature

degrees Celsius (°C)—metric unit for measuring temperature

Extension: anemometer: wind speed and direction mph; barometer: air pressure, inches; seismograph: duration and intensity of earthquakes; fathometer: sonar device to measure ocean depths; galvanometer: detects electric current; sextant: measures angular distances of stars and sun, degrees; voltmeter: electrical voltage, volts

21

MATH AT HOME

PEANUTS®

Reprinted by permission of United Feature Syndicate.

© 1988 United Feature Syndicate Inc.

YES, MA'AM.. 2½ MULTIPLIED BY 2½ IS 6¼! HOW DID I GET THE ANSWER?

HEREDITY..

MY BIG BROTHER TOLD ME!

4-8

Math: The Lighter Side

Math programs today encourage students to find answers to problems in more than one way. Knowing several ways to solve a problem helps the learner attack more difficult problems. Discuss with your child different ways that could be used to solve the problem given in the cartoon.

Multiplying Fractions

Your child is beginning a chapter on multiplication of fractions. This chapter includes problems that deal with products of fractions and whole numbers, of two fractions, and of fractions and mixed numbers. The end of the chapter begins to explore division of fractions. Share with your child ways you use multiplication of fractions, perhaps in doubling recipes.

Multiplication Pyramids

To complete each pyramid, write a number in each circle. Each number is the product of the two numbers below it. In the example:

$$\frac{3}{8} \times \frac{4}{5} = \frac{3}{10} = A$$
$$\frac{4}{5} \times \frac{5}{8} = \frac{1}{2} = B$$
$$\frac{3}{10} \times \frac{1}{2} = \frac{3}{20} = C$$

Example pyramid: C top; A, B middle; $\frac{3}{8}$, $\frac{4}{5}$, $\frac{5}{8}$ bottom.

1. Pyramid: C; A, B; $\frac{1}{4}$, $\frac{4}{5}$, $\frac{5}{6}$

2. Pyramid: C; A, B; $\frac{1}{4}$, $\frac{2}{3}$, $\frac{3}{5}$

3. Pyramid: C; A, B; $\frac{2}{3}$, $\frac{3}{4}$, $\frac{5}{9}$

Extension • Number Please!

Use number sense and what you know about fractions to solve these riddles.

$\frac{3}{5}$ of my number is 18.

$\frac{2}{3}$ of my number is 24.

The product of $1\frac{1}{2}$ times my number is 4.

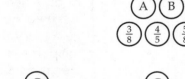

Multiplication Pyramids: 1. A = $\frac{1}{5}$, B = $\frac{2}{3}$, C = $\frac{2}{15}$; 2. A = $\frac{1}{6}$, B = $\frac{2}{5}$, C = $\frac{1}{15}$; 3. A = $\frac{1}{2}$, B = $\frac{5}{12}$, C = $\frac{5}{24}$; Extension: 30, 36, 2$\frac{2}{3}$; 22.

Volume 5   No. 23
http://www.hbschool.com

MATH AT HOME

Plane Figures and Polygons

Your child is beginning a chapter that investigates and defines plane figures and polygons. Concepts included in this chapter include line relationships, different types of angles, angle relationships, measuring angles, defining polygons, and classifying figures by angles and sides. Encourage your child to share with you activities done in class.

It's All Cut Up!

You and your child will need a paper square measuring 6 in. on each side. Cut the square into four pieces such that each piece is either a quadrilateral or a triangle. One person should mix up the pieces and give them to the other. That person will try to reassemble the square. To make it easier, on one side you could draw a picture that can be used as a guide in putting the square together. Relate this to putting together a jigsaw puzzle.

Try to make other figures and designs from your pieces.

Extension • Chop It Up!

Start with a new paper square. Cut the square into seven pieces such that each piece is either a quadrilateral or triangle. Mix up the pieces and try to reassemble the square. Was this more difficult to reassemble than the square cut into four pieces? Try to make other figures and designs from your pieces.

Math Through Children's Literature

The story line or plot of a book might include the development or use of mathematical concepts with which your child is familiar. A book might suggest an interesting mathematical investigation or pose a problem that can be solved by using mathematics.

Grand Tang's Story, by Ann Tompert (Crown Publishers, 1990) is a beautifully illustrated story that investigates the combining of geometric figures to form different pictures. The story shows how tangrams can be used to form various animals that are key characters in the story.

As a result of reading the story and seeing the changes that occur through the movement of pieces, the child should be motivated to do further independent investigations with tangrams. These experiences will help the child gain confidence working with geometric figures.

VOCABULARY

quadrilateral—a four-sided figure
parallelogram (⬭)—a quadrilateral with opposite sides equal and parallel
trapezoid (⬭)—a quadrilateral with one pair of parallel sides
rhombus (⬭)—a parallelogram with all sides equal
acute angle—an angle with a measure of less than 90°
obtuse angle—an angle with a measure of more than 90° and less than 180°

MATH AT HOME

Transformations, Congruence, and Symmetry

Your child is beginning a chapter that explores geometry in motion—changing or moving figures. Many of the ideas taught in this chapter—for example, symmetry, congruence, and transformation—can be seen often in the real world. You and your child can find examples in art, in building, in fabric designs, and around the house.

Here is an activity for you to do with your child that explores symmetry with letters. The Extension asks you to find words with mirror symmetry.

Letter Mania

When you are going to have something printed, either by a computer or at a print shop, you can choose the font, or type of print, that will be used for the letters. Which of the following alphabets, all in different fonts, have the most symmetrical letters?

1. ABCDEFGHIJKLMNOPQRSTUVWXYZ
2. ABCDEFGHIJKLMNOPQRSTUVWXYZ
3. ABCDEFGHIJKLMNOPQRSTUVWXYZ
4. **ABCDEFGHIJKLMNOPQRSTUVWXYZ**

Extension • Mirror, Mirror on the Wall

A word has mirror symmetry if the word looks like its image when seen in a mirror. For example MOM and TOT both have mirror symmetry. Can you find at least three words that have mirror symmetry? What do you notice about words with mirror symmetry?

For Your Information

Parents can help their children do well in school by supporting the work of the teachers and the school. When parents, child, and teacher work together, maximum learning can take place.

Parents should know what their child is doing in class and should check homework every day. Children should discuss with their parents the key ideas learned in school. Parents can then help their child see how these ideas are used in real life.

Children can develop their reasoning and problem-solving skills by using building materials, puzzles, and strategy games at home in connection with activities done in school. The extra practice of such games builds a student's confidence in using mathematical thinking.

VOCABULARY

transformation—the movement of a figure, either a translation, rotation, or reflection
translation—when a figure slides in any direction
reflection—when a figure is flipped across a line
rotation—when a figure is turned around a point or vertex
tessellation—a repeating pattern of closed figures that cover a surface with no gaps and no overlaps

Extension: Possible answers: TOOT, WOW, VIV; they look the same on both sides when divided down the middle.

MATH AT HOME

Circles

Your child is beginning a chapter on circles. In this chapter students will learn about the parts of a circle and the angles in a circle. They will learn how to find a circle's circumference. These ideas will then be used in making circle graphs. Challenge your child to find five different uses of circles in daily life.

Here is an activity for you and your child to do that uses circumference of a circle to make predictions in nature. The Extension investigates angles on a clock.

Hug a Tree!

You can estimate the age of a tree by measuring the distance around the trunk. The distance is called the circumference. For many kinds of trees, each 2 cm in the circumference of the trunk equals one year's growth. Find a tree and use a metric measuring tape to measure the distance around its trunk about 1 meter from the ground. That measurement represents the circumference. Divide the number of centimeters by 2 to find out how many years old the tree is. See if anyone remembers when the tree was planted.

Note: if you don't have a metric measuring tape, use a piece of string and measure the length of string that fits around the trunk with a metric ruler.

Extension • Tick Tock

Figure out how many times the hands of the clock form a right angle between midnight and noon on a given day!

For Your Information

Why are floor tiles most often hexagons or squares? Why are the letters in the word *AMBULANCE* painted in reverse on the front of the vehicle? Why are some mirrors in clothing stores set at angles?

The answers have to do with GEOMETRY. Everywhere we look we see geometry at work in the world.

Help your child see examples of shapes in your home and in nature. Point out how shapes are related and how they are used. This will help him or her develop an understanding of geometry.

Today the study of geometry is more than learning terms. It includes finding out what happens when shapes are combined, divided, and moved. Help your child see geometry as an exciting part of today's mathematics curriculum!

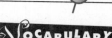

VOCABULARY

chord—a line segment that connects any two points on a circle

diameter—a chord that passes through the center of the circle

radius—a line segment that connects the center of the circle with a point on the circle

circumference—the distance around the outside of the circle (like perimeter)

MATH AT HOME

Solid Figures

Your child is beginning a chapter on solid figures. Problems in this chapter involve working with prisms and pyramids; looking at relationships among the faces, edges, and vertices of a solid figure; picturing solids and using nets (patterns) to make them; and solving problems about volume. Ask your child to share with you problems done in class.

Here are an activity and an Extension that involve picturing different views of a cube.

Cube It!

This is a pattern that, when folded, will make a cube.

It is not the only net, or pattern, that will make a cube. Draw at least four more different patterns that will also make a cube. How are your nets alike? How are they different?

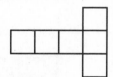

Extension • All Sides Accounted For

Look at the different views shown of the cube below. Based on these views, figure out which letters are on opposite faces of the cube. Make a model of the cube if you need to.

A is opposite _____
T is opposite _____
S is opposite _____

For Your Information

When you hear the word volume, what comes to your mind? Do you think of the loudness of a TV or of how much a container holds? Many words have more than one meaning.

Children will meet many, many new words in their study of mathematics. Some of these words are also used outside of math, and some are used for mathematics alone. It is important that children be able to use the vocabulary of mathematics.

Understanding the vocabulary is a key to success in problem solving and in other operations. Children should know how to use the glossary in their book to find the meanings of words. They should take time to rewrite definitions in their own words to help them remember what the math words mean. Vocabulary is an important part of mathematics.

VOCABULARY

pyramid—a solid figure with one base that is a polygon and three or more faces that are triangles

prism—a solid figure that has two congruent parallel faces called bases

face—a flat surface of a solid figure

vertex—the point where three or more edges of a solid figure meet (plural—**vertices**)

net—a two-dimensional pattern for a three-dimensional solid figure

MATH AT HOME

PEANUTS®

SORRY, MA'AM..I THINK I WAS ON CRUISE CONTROL

Math: The Lighter Side

Success in learning requires that the learner take an active part. In today's math classes, children use mathematical models, tools, and technology to actively involve themselves in their learning. Group work, such as projects and problem solving, provides still another way to have the children take an active part. It is also important that children use their learning outside the classroom.

Fractions as Ratios

Your child is beginning a chapter that teaches the math idea of ratio. Problems in the chapter include expressing ratios, writing equivalent ratios, and using ratios in connections with probability and map scales. Share with your child ways you may use ratios in daily life.

A Popping Experience

Count the number of kernels in a half-cup of unpopped popcorn kernels. Also find the weight of the half-cup of unpopped kernels. Pop the kernels. Count the number of kernels that popped. Weigh a half-cup of the popped kernels. Express the ratio of unpopped kernels to popped kernels. Repeat for different amounts of kernels. Were your results the same as the first time?

Extension • Batting Champ!

A batting average is the ratio of the number of hits to the number of times at bat. Who has the best batting average in the list at the top of column two?

Player 1: 14 hits out of 50 times at bat

Player 2: 67 hits out of 250 times at bat

Player 3: 54 hits out of 200 times at bat

TESTING Tip

Remind your child that answers to test questions must be in the form that was asked for.

VOCABULARY

ratio—two numbers that compare a part with a whole, a whole with a part, or a part with a part

map scale—a ratio that compares the distance on a map with the actual distance

similar—figures that have the same shape, but are not necessarily the same size

Extension: Player 1

Volume 5   No. 28
http://www.hbschool.com

MATH AT HOME

Percent

Your child is beginning a chapter that teaches the math idea of percent. Problems in the chapter focus on the meaning of percent; equivalence of fractions, decimals and percents; using percent benchmarks, and finding percents of a number. Share with your child some of the ways you use percents in daily life.

Here is an activity that shows how characters are made on a computer screen. The Extension asks you and your child to discuss whether given statements about percents make sense.

Letters in Lights

Characters on computer screens are made by lighting dots (pixels) on a square or rectangle. Make a 10×10 square grid to serve as a pixel on which you can write letters of the alphabet that meet the given conditions.

a. Make a letter of the alphabet that uses 50% of the squares in the pixel.

b. Make a letter of the alphabet that uses more than 75% of the squares in the pixel.

c. Make a letter of the alphabet that uses less than 25% of the squares in the pixel.

Extension • A Matter of Opinion

Discuss whether the following statements make sense.

John says he is right 100% of the time.

Mary wants to spend 100% of her allowance on candy.

A championship basketball team makes only 10% of its shots.

Even if Mary misses 10 problems on a test, she can still get an A.

Math in the News

What do interest rates, scales, and basketball standings have in common? They all involve percent. Percents have a great many uses in daily life. Buyers should know the meaning of percent and how to solve problems involving percent. These problems might include finding the percent of a number, the percent one number is of another, or a number when you know a percent of it. Many percents can be easily found, for example, 25% by taking $\frac{1}{4}$ of a number, 50% by taking $\frac{1}{2}$ of a number, and 10% by moving the decimal point one place to the left.

Whether you are figuring out discounts, tips, takes, or rankings, you will find yourself 100% involved with percent!

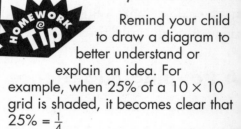

HOMEWORK Tip   Remind your child to draw a diagram to better understand or explain an idea. For example, when 25% of a 10×10 grid is shaded, it becomes clear that $25\% = \frac{1}{4}$

VOCABULARY

percent—a ratio of some number to 100
benchmark percent—a commonly used percent that is close to the amount you are estimating

Extension: It is not possible for John to be right 100% of the time. It is possible that Mary will spend all of her allowance. The team makes 1 out of every 10 shots. Mary has an A average.

28